Princess Wendy was as lovely as the sweetest rose. There was just one thing. Wendy had the longest feet in Hamlet Kingdom.

"Try sneakers and longer woolen socks," Queen Mandy said. "They could make your feet look shorter."

"Not likely," answered Wendy. "And I am thankful for my feet. Who wants elfish little footlets when going for a run?"

Still, Queen Mandy was upset. Who would wed a princess with the longest feet in Hamlet?

This did not upset Wendy a bit. She just kept on going. She ran endless miles, streaking by trees and leafy bushes.

"Hardy feet are handy," Wendy said.

Then one day, Wendy met Mr. Peerless.

"Are you a tourist?" Wendy asked.

"I am a coach," Mr. Peerless said. "I am teaching people how to play soccer. It is not dangerous."

"Please teach me," pleaded Wendy. "I really want to play!"

"Kick that," Coach Peerless said.

So Wendy kicked.

"My goodness!" Coach Peerless yelled. "With your help, Hamlet can win the game tomorrow!"

Hamlet really did win. Wendy made more goals than any soccer player in the world!

So was Queen Mandy happy then? Well, yes.

Coach Peerless fell in love with Wendy and asked her to be his bride.